HOW TO GROW PUMPKIN

Beginners guide to growing, caring and harvesting of pumpkin from seed

Larry Pat

TABLE OF CONTENT

INTRODUCTION

Pumpkin cultivation is an ancient and fascinating practice that spans cultures and continents, offering a rich tapestry of agricultural history and culinary tradition. The pumpkin, scientifically known as Cucurbita pepo, is a warm-season crop that belongs to the gourd family, Cucurbitaceae. Renowned for its vibrant orange hue, robust flavor, and versatile culinary applications, the pumpkin has become an iconic symbol of autumn and harvest festivals worldwide.

The origins of pumpkin cultivation can be traced back to Mesoamerica, where indigenous peoples such as the Aztecs and Mayans first domesticated wild varieties of squash around 5500 B.C. These early cultivators selectively bred plants for desirable traits, gradually developing the diverse range of pumpkin varieties we know today. The word "pumpkin" itself has its roots in the Greek word "pepon," meaning "large melon."

Pumpkins are characterized by their sprawling vines, large lobed leaves, and strikingly large fruits, which can range from small, sugar pumpkins to massive varieties commonly used for carving during Halloween. Cultivating pumpkins involves careful consideration of soil conditions, climate, and cultivation practices, making it both an art and a science.

Chapter 1: Introduction to pumpkin cultivation

Overview of Pumpkin Cultivation

Pumpkins, belonging to diverse species within the Cucurbita genus, stand out as a significant reservoir of carotenoids, playing a crucial role in human nutrition. Predominantly, β-carotene emerges as the primary carotenoid, exhibiting concentrations exceeding 70 µg/g across various species. However, noteworthy variations in carotenoid content manifest not only between different species but also within the same species or variety. This variability is attributed to a medley of factors, including the nuances of growing conditions, the maturation stage of the pumpkins, and the intricacies of harvesting and post-harvesting treatments.

Compounding the intricacies, the preparation and processing of pumpkins introduce an additional layer of complexity. During these stages, the potential for oxidation and/or isomerization of carotenoid compounds exists, leading to a consequential reduction in biological activity and potential color loss. The intricacies of these reactions are intricately linked to factors such as temperature fluctuations, exposure to oxygen, and the influence of light.

Historical and Cultural Significances

Pumpkins have a rich historical significance in festivals and celebrations, originating from indigenous cultures in the Americas. These cultures, including the Algonquian, Iroquois, and Cherokee tribes, integrated pumpkins into their traditional agriculture and cuisine, using them as a staple food source and incorporating them into cultural practices and rituals.

The pumpkin held a special place in the agricultural calendars and spiritual beliefs of these indigenous communities.

With the colonization of the Americas, European settlers recognized the versatility and nutritional value of pumpkins. They adopted indigenous cultivation techniques and incorporated pumpkins into their culinary traditions, leading to the integration of pumpkins into both savory and sweet dishes in the evolving cuisines of the American colonies.

Pumpkins' cultural significance expanded as festivals and celebrations became intertwined with the agricultural cycles of early American communities. Harvest festivals, in particular, celebrated the land's bounty, with pumpkins symbolizing abundance and sustenance.

The tradition of carving pumpkins, later associated with Halloween, has its roots in Celtic and European folklore, where hollowed-out turnips and later pumpkins were used as lanterns to ward off evil spirits.

As the United States expanded westward, pumpkins remained integral to agricultural communities, deeply ingrained in American society. The enduring cultural significance of pumpkins is evident in traditions surrounding pumpkin festivals, culinary customs, and the widespread use of pumpkins as decorative elements in seasonal celebrations.

Chapter 2: The World of Pumpkins

Botanical Classification and Varieties

The pumpkin belongs to the plant genus Cucurbita and the family Cucurbitaceae. There are several species within the genus Cucurbita that are commonly referred to as pumpkins, including Cucurbita pepo, Cucurbita maxima, Cucurbita moschata, and Cucurbita mixta. These species encompass a wide variety of pumpkin varieties and cultivars.

1. Cucurbita pepo: This species includes many small- to medium-sized pumpkins and squashes. Examples of pumpkins within this species include the classic Jack-o'-lantern pumpkins, acorn squash, and zucchini.

2. Cucurbita maxima: This species includes larger pumpkins, often with thick, orange flesh. Varieties within this species include the Atlantic Giant pumpkin and Hubbard squash.

3. Cucurbita moschata: This species includes pumpkins with tan or greenish skin and sweet, orange flesh. Butternut squash is a well-known variety within this species.

4. Cucurbita mixta: This species includes pumpkins that don't fit clearly into the other three species. One example is the Cushaw pumpkin.

Botanically, pumpkins are dicotyledonous plants, meaning they have two seed leaves (cotyledons) upon germination. They belong to the order Cucurbitales, which also includes other gourd family members like cucumbers and melons.

1. Cooking and Baking

Pumpkin Pie: Perhaps the most iconic use, pumpkin pie is a classic dessert during the fall and Thanksgiving season.
Pumpkin Soups and Stews: Pumpkin's natural sweetness and creamy texture make it an excellent base for soups and stews.
Roasted Pumpkin: Cubes of pumpkin can be roasted with herbs and spices for a delicious side dish.

2. Breads and Muffins

Pumpkin Bread and Muffins: Pumpkin puree can be added to bread and muffin recipes for moisture and flavor.

3. Pumpkin Seeds

Roasted Seeds: Pumpkin seeds, also known as pepitas, are often roasted and seasoned as a nutritious snack.

4. Pumpkin Puree

Ingredient in Various Dishes: Pumpkin puree can be used as an ingredient in a wide range of dishes, from pasta sauces to smoothies.

5. Pumpkin Beverages

Pumpkin Spice Latte: The famous seasonal coffee drink incorporates pumpkin flavor with spices.

Pumpkin Smoothies: Pumpkin puree can be added to smoothies for a fall twist.

6. Pumpkin as a Side Dish

Mashed Pumpkin: Similar to mashed potatoes, pumpkin can be boiled and mashed for a side dish.

Pumpkin Fries: Sliced pumpkin can be baked or fried as a tasty alternative to traditional fries.

1. Carved Jack-o'-Lanterns

Halloween Decor: Hollowed-out pumpkins are carved into various designs, creating spooky or whimsical decorations for Halloween.

2. Autumn Centerpieces

Table Decor: Whole pumpkins or arrangements of small pumpkins and gourds are popular as centerpieces for fall and Thanksgiving tables.

3. Pumpkin Displays

Outdoor Decor: Pumpkins of different sizes, shapes, and colors can be arranged on porches or in gardens to create visually appealing displays.

4. Fall Festivals and Events

Decorative Competitions: Pumpkins are often featured in fall festivals, with competitions for the largest, most creatively carved, or uniquely decorated pumpkins.

5. Crafts and DIY Projects

Painted Pumpkins: Pumpkins can be painted with various designs for a creative and longer-lasting decoration.
Pumpkin Wreaths: Dried pumpkin slices or artificial pumpkins are used in crafting decorative wreaths.

6. Symbol of Fall Harvest

Symbolism: Pumpkins are symbolic of the fall harvest season and are often used to evoke a sense of warmth and abundance.

Chapter 3. Environmental Considerations

Climate and Geographical Preferences

Pumpkins thrive in warm climates and are sensitive to frost. Here are some considerations regarding climate and geographical preferences:

1. Temperature

Warm Season Crop Pumpkins are warm-season crops, and they require a frost-free growing season. The ideal temperature range for pumpkin cultivation is between 70°F and 95°F (21°C to 35°C).

Growing Season: Pumpkins need a relatively long growing season, typically around 75 to 100 frost-free days.

2. Sunlight

Full Sun: Pumpkins prefer full sunlight exposure for at least 6 to 8 hours a day. Adequate sunlight is crucial for the development of healthy plants and the production of quality fruit.

3. Geographical Considerations

Growing Zones: While pumpkins can be grown in various climates, they are often associated with regions that have warm summers. They can be successfully cultivated in USDA Hardiness Zones 3-9.

Soil Requirements and Preparation

Pumpkins are relatively adaptable when it comes to soil types, but certain conditions are more favorable for optimal growth:

1. Soil Type

Well-Drained Soil: Pumpkins prefer well-drained, loamy soil that is rich in organic matter. Good drainage helps prevent waterlogging, which can lead to root diseases.

pH Level: The ideal soil pH for pumpkins is slightly acidic to neutral, ranging from 6.0 to 7.5.

2. Soil Preparation

Amending Soil: Before planting, it's beneficial to amend the soil with organic matter, such as compost, to improve fertility and structure.

Raised Beds: In areas with heavy or poorly-draining soil, growing pumpkins in raised beds can provide better drainage.

3. Moisture Considerations

Consistent Moisture: Pumpkins require consistent moisture, especially during the flowering and fruiting stages. Mulching around the plants aids in retaining soil moisture and suppressing weeds.

Avoid Waterlogged Soil: While pumpkins need adequate water, it's important to avoid waterlogged conditions, as this can lead to root rot.

4. Companion Planting:

Companion Plants: Planting pumpkins with companion plants like corn and beans (the Three Sisters planting method) can be beneficial. Corn provides support for climbing pumpkin vines, and beans contribute nitrogen to the soil.

5. Crop Rotation

Rotation: Avoid planting pumpkins in the same location in consecutive years to reduce the risk of soil-borne diseases.

Chapter 4. Getting Started

Seed Selection and Acquisition

Ah, the thrill of starting a pumpkin patch! Picture this: you, the cultivator of autumnal wonders, carefully selecting the seeds that will soon burgeon into vibrant, plump pumpkins. The first step is choosing the right seeds:

Venture into the realm of seed catalogs or your local garden center, and let your imagination roam. Will it be the classic Jack-o'-lantern pumpkin, a whimsical heirloom variety, or perhaps a petite pie pumpkin? Consider your culinary dreams and decorative aspirations as you pick seeds that align with your pumpkin vision.

Planting Techniques: Direct Sowing vs. Indoor Start

Now, the canvas awaits your green thumb. The decision: direct sowing or the enchantment of indoor starts?

Direct sowing, the dance with nature: Imagine your garden as a blank canvas, the earth beneath your fingers. With the warmth of the sun as your ally, sow your pumpkin seeds directly into the soil when the frosty grip of winter has loosened. Watch as the tiny green heralds emerge, each one a promise of autumn treasures.

Or, the indoor start, a nurturing embrace: Picture a cozy corner bathed in soft light. This is your pumpkin nursery. Start your seeds indoors, providing them a head start in the tender care of pots and window sills.
When the time is ripe, transplant these seedling champions into the garden, ensuring a robust and early harvest.

Germination and Seedling Care

As the days pass, anticipation brews. Behold the magic of germination!

Patience is the virtue of the pumpkin grower. Witness the tiny cotyledons unfurl, delicate green sails navigating the sea of soil. Keep the soil consistently moist but not drenched, offering a nurturing environment for your fledgling pumpkin ambassadors.

As they grow, shower them with tender care. Shield them from unexpected chills, and ensure they bask in the sun's gentle embrace. Mulch around their tender stems to conserve moisture and fend off unruly weeds. These are the early days of a partnership that will yield autumnal splendor.

Chapters 5: Nurturing Healthy Plants

Watering Practices

Ah, the elixir of life for your burgeoning pumpkin kingdom — water. Imagine yourself as the steward of a verdant realm, entrusted with the task of quenching the thirst of your pumpkin progeny.

Water wisely, my horticultural companion. Provide a consistent and even supply, allowing the soil to remain consistently moist but not waterlogged. The thirsty tendrils of your pumpkin vines crave hydration, particularly during dry spells. A deep, thorough watering is your gift to ensure robust growth and succulent fruit.

Morning showers or evening serenades? Consider watering in the morning to give your pumpkin patch time to dry before the cool evening. This helps prevent fungal diseases and encourages a harmonious dance between sunlight and droplets on leaves.

Fertilization Guidelines

Behold the feast laid out for your pumpkin progeny! Fertilization, the nourishing banquet that propels your vines toward splendiferous vitality.

Choose your elixirs wisely. Opt for a balanced, all-purpose fertilizer or one specifically formulated for vegetables. Begin the feast in the early stages, providing nutrients to fortify the roots and fuel the climb skyward. As the first blossoms grace your vines, shift to a fertilizer higher in phosphorus to encourage robust fruiting.

Feast in moderation. While generosity is noble, avoid overfeeding. A balanced approach prevents lush foliage at the expense of fruit development. Your goal is a harmonious equilibrium, a nourishment symphony for your burgeoning pumpkin ensemble.

Importance of Proper Spacing

Ah, the dance floor of the pumpkin patch — where vines twirl, leaves pirouette, and pumpkins stretch their limbs. Proper spacing, the choreography that ensures each performer has room to shine.

Grant each pumpkin its stage. Avoid overcrowding, for a crowded stage leads to tangled vines and stifled growth. Follow the spacing recommendations for your chosen pumpkin variety, allowing for ample airflow and sunlight penetration.

The art of distancing. Adequate spacing wards off the specter of diseases, as it prevents the unwelcome embrace of moisture-laden shadows. Each pumpkin, with its sprawling leaves, deserves its own patch of sunlight to thrive.

As you tend to the water, offer the elixirs of fertilization, and choreograph the dance of spacing, envision your pumpkin patch as a living tapestry, a lush testament to the artistry of nature and your nurturing hand. The fruits of your labor will be more than pumpkins; they will be a harvest of fulfillment and the embodiment of a gardener's enchanting touch.

Chapter 6: Challenges in Pumpkin Cultivation

Common Pests and Diseases

In the whimsical dance of cultivating pumpkins, challenges may emerge, casting shadows on the vibrant pumpkin patch. Picture this: tiny invaders and unseen maladies threatening your orange-hued dreams.

1. Cucumber Beetles: These cunning creatures can transmit bacterial wilt and munch on leaves.

2. Squash Bugs: Stealthy and destructive, they pierce plant tissue and suck out the sap.

3. Vine Borers: The silent infiltrators, larvae bore into stems, causing wilting and eventual collapse.

4. Aphids: Minuscule troublemakers, they cluster on leaves, extracting plant juices.

5. Powdery Mildew: A spectral veil on leaves, hindering photosynthesis and weakening the plant.

6. Downy Mildew: A stealthy foe, spreading its mold-like growth on the undersides of leaves.

7. Bacterial Wilt: The consequence of beetle attacks, causing wilting and eventual demise.

8. Anthracnose: Dark lesions on fruits and foliage, a harbinger of rot.

Integrated Pest Management Strategies

Fear not, valiant gardener! Arm yourself with the tools of integrated pest management, a shield against the encroaching forces. Envision your garden as a fortress, and these strategies as your stalwart defenders:

Companion Planting
- Marigolds: Plant these golden sentinels to repel nematodes and deter pests.
- Nasturtiums: Their peppery allure may deter aphids and squash bugs.
-

Beneficial Insects
- Ladybugs: Nature's tiny warriors, feasting on aphids with voracious appetites.
- Parasitic Wasps: Allies in the battle against vine borers and other caterpillar pests.

Cultural Practices

- Crop Rotation: Baffle pests with a strategic shuffle of your pumpkin patch layout each season.
- Timely Harvest: Swiftly gather mature fruits to thwart the spread of diseases.

Organic Sprays

- Neem Oil: A botanical guardian, it disrupts the life cycle of pests and acts as a fungicide.
- Garlic Spray: Ward off vampires and pests alike; a pungent deterrent.

Vigilance and Early Intervention

Scout regularly: A vigilant watch prevents the subtle advances of pests and diseases.

Prune affected foliage: Swift removal of infected leaves curtails the spread.

Chapter 7: The Growth Cycle

Flowering and Pollination

Imagine your pumpkin patch as a vibrant orchestra, each plant a musician contributing to the grand symphony of growth. As the season unfolds, the overture begins with the delicate notes of flowering and pollination.

The Blossoming Ballet
- Floral Elegance: Picture the emergence of bright yellow flowers, the heralds of impending pumpkin glory.
- Dance of the Bees: Envision busy bees, nature's pollination maestros, flitting from flower to flower, transferring golden dust.

The Pollination Waltz
- Male and Female Blooms: In the pumpkin kingdom, separate genders reign. Male blooms

appear first, followed by the regal female blooms.

- The Bee's Embrace: Watch as pollen journeys from male to female blooms, setting the stage for fruit formation.

Fruit Development

In the heart of your garden's symphony, the crescendo builds with the development of your prized pumpkin fruits. Picture the gradual transformation from floral elegance to burgeoning fruit.

The Infant Gourd

- Tiny Protuberance: Envision the miniature pumpkin, a mere swell on the vine, promising potential.
- Growth Spurt: Marvel as your pumpkin undergoes a growth spurt, expanding into a recognizable form.

The Maturation Sonata

- Deepening Hues Picture the evolving colors, from muted greens to the rich, vibrant orange of maturity.
- Fruit Setting and Swelling: Observe the swelling of your pumpkin, a testament to successful pollination and impending harvest.

Monitoring Growth Stages

As the garden symphony reaches its zenith, take on the role of conductor, monitoring the nuanced stages of growth with an attentive eye.

Visual Indicators

- Vine Tendrils: Note the drying and withering of vine tendrils near the pumpkin stem, a sign of ripeness.
- Hardening Rind: Gently tap the pumpkin; a mature fruit will have a hard, unyielding rind.

- Deep Coloration: A rich, consistent color signals the readiness for harvest.

Timing the Harvest Ballet
- Late Summer to Fall: Envision the harvest window, typically from late summer into fall.
- Harvest Ritual: Picture the joyous moment as you delicately sever the pumpkin from its vine, celebrating the culmination of your gardening journey.

Chapter 8: Harvesting and Post-Harvest Handling for Pumpkins

Determining Pumpkin Harvest Readiness

Pumpkins, like many crops, have specific indicators of readiness for harvest. A key factor is the pumpkin's color—ripe pumpkins typically exhibit a deep, consistent coloration. Additionally, the stem connecting the pumpkin to the vine should be dry and hard. Tapping the pumpkin should produce a hollow sound, signaling maturity. Regular monitoring and familiarity with the pumpkin variety's characteristics aid in accurate harvest assessments.

Pumpkin Harvesting Techniques

The harvesting technique employed for pumpkins depends on the scale of cultivation. For small-scale or home gardens, hand harvesting is common. It involves cutting the stem, leaving a few inches attached to the pumpkin. For larger-scale operations, machine harvesting can be efficient, with equipment designed to gently detach pumpkins from the vine. Care should be taken to avoid bruising or damaging the pumpkins during the harvesting process.

Pumpkin Storage and Preservation Tips

Proper storage is crucial to preserving the freshness of harvested pumpkins. Store pumpkins in a cool, dry place with good ventilation to prevent mold and decay. Avoid placing them directly on the ground, using pallets or racks to allow air circulation.

Inspect pumpkins for any signs of damage or disease before storage. Preservation techniques for pumpkins include curing for a few weeks in a warm, dry area to toughen the skin and enhance flavor. For longer storage, consider canning, freezing, or dehydrating pumpkin products.

Chapter 8: Delicious Pumpkin Recipes

1. Pumpkin Spice Latte

Ingredients: Pumpkin puree, espresso, milk, pumpkin spice, whipped cream.

Instructions: Combine pumpkin puree, espresso, and milk, add pumpkin spice, and top with whipped cream.

2. Roasted Pumpkin Salad

Ingredients: Roasted pumpkin cubes, mixed greens, feta cheese, pumpkin seeds, balsamic vinaigrette.

Instructions: Toss roasted pumpkin with greens, feta, and pumpkin seeds. Drizzle with balsamic vinaigrette.

3. Pumpkin and Sage Risotto

Ingredients: Arborio rice, pumpkin, sage, onion, garlic, Parmesan cheese, vegetable broth.

Instructions: Sauté onion and garlic, add rice, pumpkin, and sage. Gradually add broth and Parmesan until creamy.

4. Pumpkin Hummus

Ingredients: Chickpeas, pumpkin puree, tahini, lemon juice, garlic, cumin, olive oil.

Instructions: Blend chickpeas, pumpkin, tahini, lemon juice, garlic, and cumin until smooth. Drizzle with olive oil.

5. Pumpkin and Black Bean Quesadillas

Ingredients: Pumpkin puree, black beans, shredded cheese, tortillas, cumin, chili powder.

Instructions: Spread pumpkin on tortillas, add black beans, cheese, cumin, and chili powder. Grill until the cheese melts.

6. Creamy Pumpkin Soup

Ingredients: Pumpkin, onion, garlic, vegetable broth, coconut milk, nutmeg.

Instructions: Sauté onion and garlic, add pumpkin, broth, and coconut milk. Simmer, then blend until smooth. Season with nutmeg.

7. Pumpkin and Spinach Stuffed Shells

Ingredients: Jumbo pasta shells, pumpkin ricotta filling, marinara sauce, mozzarella cheese.

Instructions: Stuff shells with a mix of pumpkin and ricotta, bake with marinara and mozzarella until bubbly.

8. Pumpkin and Chocolate Swirl Brownies

Ingredients: Pumpkin puree, brownie mix, chocolate chips.

Instructions: Mix pumpkin into brownie batter, swirl with chocolate chips, and bake as directed.

9. Pumpkin and Caramel Cheesecake

Ingredients: Graham cracker crust, cream cheese, sugar, eggs, pumpkin puree, caramel sauce.

Instructions: Mix cream cheese, sugar, eggs, and pumpkin. Swirl in caramel and bake in the crust.

10. Pumpkin Spice Pancakes

Ingredients: Pancake mix, pumpkin puree, milk, pumpkin spice.

Instructions: Mix pancake ingredients, adding pumpkin spice. Heat a pan over medium heat. Put 1/4 cup of batter for every pancake. Cook until bubbles form on top, then flip and cook the other side until golden brown.

CONCLUSION

The cultivation of pumpkins is a rewarding and versatile endeavor that extends beyond the mere act of growing a seasonal vegetable. Beyond their vibrant orange hue and robust flavor, pumpkins serve as a symbol of autumnal abundance and tradition.

The cultivation process, from seed to harvest, fosters a deeper connection to the rhythms of nature and the agricultural cycle. Whether grown in a backyard garden or on a larger scale, pumpkins contribute not only to culinary delights but also to festive decorations, carving traditions, and a sense of community during fall celebrations.

With their adaptability to various climates and relatively straightforward cultivation requirements,
pumpkins stand as an accessible and fulfilling option for both novice and

seasoned gardeners alike. As we reflect on the cultivation journey, it becomes evident that growing pumpkins goes beyond horticulture—it is a celebration of the seasons, an invitation to connect with the earth, and a source of joy for those who partake in the wholesome process of nurturing and harvesting these iconic gourds.